# Intelligent Compaction of Soils and Subbase Materials:
## A Broad Overview

I0481829

Written by:
## Christopher Wanamaker, P.E.

May 18th, 2016

# Table of Contents

# Introduction

The key to constructing any high performing foundation or pavement system is having a strong base that can support the structure being placed on it. Proper compaction of base and subbase materials is probably the single most important thing that can be done to improve the soil's bearing capacity thus ensuring a stabile and long lasting final product. Traditionally, compaction of soils and subbase materials has been mostly achieved using heavy rollers that make multiple passes over the surface. Good compaction is achieved by compacting the structural section in sequential layers. Quality control methods have included the use of a variety of specialized equipment designed to measure in-situ soil density and moisture content at specified locations.

The current practice of compaction and quality testing leaves much room for error. Uniformity of compaction is often the goal however it is rarely achieved. Testing compaction in a regular pattern or at random still leaves most of the project site untested or underrepresented. Project managers rely on these tests, in conjunction with statistical measures, to provide representative quality indicators for the entire project site. Problem areas of low quality compaction can still exist within the project site that may be overseen or passed over when relying on standard point density tests [3]. Increasing the amount of test location can improve quality however this can often come with significant cost.

When a smooth drum or sheepsfoot roller is equipped with intelligent compaction technology, the onboard computer uses the feedback it receives from its sensors to make adjustments to compactive effort. The onboard computer achieves this by modifying the amplitude and frequency of the roller vibrations thus changing the amount of force imparted on the soil or asphalt layers [4]. For some machines, the speed of the roller can also be modified which results in a change in the amount of energy per distance unit that is imparted on the soil. The onboard computer and software code, coupled with an auto-feedback mechanism, is essentially what allows the machine to adapt its behavior to the changing site conditions [1]. In essence, this is what differentiates intelligent compaction from older sensor-based or computer assisted compaction technologies.

The goal of intelligent compaction is to address the quality control gaps in the current practice. Furthermore, intelligent compaction serves to provide a means to consistently and instantaneously provide the project manager or engineer with compaction information all throughout the entire construction process. This can reduce, or potentially eliminate, the need for spot testing and helps to ensure a consistent and uniform amount of compaction throughout the entire project site. Moreover, the data generated with intelligent compaction equipment can be used in conjunction with other quality control measures to ensure that a quality product is being constructed. All of these things lead to a reduction in rework, improved efficiency, a lower cost for construction [9], and longer lasting roadway. In addition to this, over-compaction is avoided as the computer automatically adjusts the compaction process to prevent compacting an area that has already met the required stiffness level [10].

# __Background__

In 1974 Dr. Heinz Thurner, a Swedish Highway Administration official, introduced the first continuous compaction controlled (CCC) roller design when he attached an accelerometer to a roller compactor. This design measured a unitless number which he called the compaction meter value (CMV). This number was calibrated against standard density tests at the time and was used to help improve compaction consistency and quality. However, without the aid of a reliable onboard computer to make automatic adjustments to the compactive effort, the operator of the roller would need to read the output data from the device and manually make adjustments. The roller operator had to control many variables such as the compaction speed, vibration frequency, and the number of passes. A skilled roller compactor driver could make these adjustments quickly with relatively good accuracy. This basic design, called continuous compaction control (CCC) was commercialized in 1980 [2] by Geodynamik. Manufacturers of these rollers have made incremental improvements in the technology ever since CCC's inception. Intelligent compaction was born when the standard continuous compaction control system was paired with an onboard auto-feedback computer capable of changing the vibrating drum's frequency, amplitude, and speed to improve compaction in real time.

Europe and Japan were the first countries to make widespread use of intelligent compaction [3]. Around 2005 the Federal Highway Administration (FHWA) began funding research and development of Intelligent Compaction Technology in the United States. Soon afterwards, the FHWA partnered with 12 state departments of transportation and commission pilot projects in an effort to expand the knowledge and use of intelligent compaction technology. In 2013 the FHWA selected intelligent compaction for inclusion in its "Every Day Counts" initiative which served to focus even more resources on improving intelligent compaction while also making efforts to introduce it at a national scale [2].

In addition to the slow adoption of intelligent compaction, there has been an ongoing shift among state agencies to get away from the standard quality control methods which rely on point density measurements and move towards a measurement of soil/pavement stiffness values [4], [6] instead. In 1933, R.R. Proctor developed the standard Proctor test that is currently in use for determining the maximum dry density of a soil at an optimum moisture content. Quality assurance during construction is typically achieved by measuring density and moisture content in the field and comparing it to the laboratory (Proctor) values. One of the major disadvantages to this approach has been that the density of a particular soil does not necessarily correlate to the strength of the soil because strength is influenced by a variety of factors besides density and moisture content. In addition to this, roadway and pavement structures aren't designed with density data. In fact, a roadway designer will use a stiffness (modulus) value instead [4], [6].

# Intelligent Compaction Technology

Intelligent compaction by definition means the use of "smart" vibratory rollers to measure compaction in real time. Modern intelligent rollers are equipped with a global positioning systems (GPS), accelerometers, and a variety of other sensors to measure compaction, compaction efficiency, physical location on the job site, and even the number of passes. An onboard computer monitors the roller's travel speed as well as compactive effort and makes constant adjustments to the drum's vibration frequency and amplitude. These calculated adjustments help to ensure that the compaction of the soil is as effective and efficient as possible. The result is a more consistently compacted and higher quality soil or pavement structure.

The global positioning system (GPS) is a key piece of technology needed to make the intelligent compaction system work. The typical GPS unit installed on a compactor utilizes a real-time kinematic (RTK) type of system in order to measure the exact location of the roller within about 0.98 inches horizontally and 1.46 inches vertically [1]. The RTK system is much more accurate than older satellite differential GPS units. With that said, many of this systems require a base station to be set up at the project site in order to maintain constant survey control throughout the duration of construction. The GPS unit is used to generate geospatial data, monitor roller speed, and verify the number of passes already completed.

Another key component of the intelligent compaction system is the accelerometer transducers. The accelerometers are the devices that actually measure movement of the vibratory drum which is then correlated to a value for the soil's stiffness using an equation specific to the manufacturer of the compaction roller. There are typically two different ways that the accelerometers are utilized: 1) To measure the ratio of frequency changes of the drum's response to the soil stiffness over set time intervals or 2) To calculate soil layer stiffness directly based on drum-ground interactions [3].

Another way that compaction can be measured is by carefully monitoring rolling resistance, engine efficiency, and/or machine drive power. The greater the resistance to rolling motion the less compact a soil or pavement layer is because more power is needed to drive the roller forward. Typically, these types of measurements work best when paired with the accelerometers located on the vibratory drum.

All measured data is typically taken at 0.3 to 1.6 feet (0.1 to 0.5 meter) intervals and assigned GPS coordinates when stored in the computer [9]. This data can be used in geographic information system (GIS) environment for review, verification, and permanent storage. With this information, quality assurance can occur in near real time. Furthermore, the wealth of information generated by the intelligent compactor can be used for future studies, improvements to rolling and paving methodologies, as well as the evaluation of roadway longevity.

# Intelligent Compaction Equipment

Intelligent compaction rollers have been designed to compact just about any type of material that requires compaction. In particular, there are five types of materials that typically need to be compacted for construction purposes [3]:

- Type I - Granular, Non-cohesive subgrade soils
- Type II - Fine-grained, cohesive subgrade soils
- Type III - Aggregate base material
- Type IV - Asphalt pavement material
- Type V - Stabilized Subbase material

Intelligent compaction enabled rollers are capable of handling any of these material types. In the case of cohesive soils, the roller will be equipped with a sheepsfoot or padfoot drum in order ensure that compaction will be effective. Most granular soils are compacted with smooth drum rollers. Single drum rollers are used to compact soils and dual or double drum rollers are used to compact asphalt pavements.

Currently, there are 5 global manufacturers which produce intelligent compaction enabled vibratory rollers for use in compaction of soil: Bomag, Ammann/Case,

Caterpillar, Dynapac, and Sakai [1]. These manufacturers produce both smooth wheel and sheepsfoot or padfoot wheel rollers. Trimble, a leading producer of small construction equipment, also manufactures and sells kits that can be used to retrofit many existing rollers into an intelligent compaction enabled device. Each manufacturer of intelligent compaction rollers utilizes the same basic setup in terms of equipment which includes GPS receivers, accelerometers, a computer, and an infrared camera (only for use in asphalt compaction). However, these manufacturers use different software, user interfaces, as well as different calibration techniques and compaction/stiffness measurement methodologies. A description of each manufacturer's methodology is discussed next.

### *Bomag Intelligent Compaction Equipment*
Bomag's intelligent compaction technology is called "VarioControl." This manufacturer uses two accelerometers to directly measure a vibration modulus, $E_{vib}$, (stiffness value). The computer uses information about the ground contact force and roller displacement to calculate this modulus value. This system has a feedback feature that responds to the soil's stiffness by varying the vibration amplitude to improve compaction efficiency [1], [7]. The

vibration modulus (MN/m$^2$) is solved iteratively using the following equation:

$$\frac{\Delta F}{\Delta z_1} = \frac{2E_{vib}a\pi}{2(1-v^2)\left[2.14 + 0.5\ln\left(\frac{\pi(2a)^3 E_{vib}}{16(1-v^2)(m_b + m_e + m_r)g(\frac{d}{2})}\right)\right]}$$

Where the ratio of the change in applied force over the roller displacement is equivalent to the drum dimensions, soil vibration modulus, and Poisson's ratio for the soil [3].

### *Ammann/Case Intelligent Compaction Equipment*

This manufacturer's proprietary system is called the "ACEplus" system which combines their Ammann Compaction Expert (ACE) measurement values with a new intelligent compaction control system. The system measures a lumped soil stiffness value called the $k_b$ value which represents the spring-like interaction between the vibrating drum and the soil's surface. ACE uses an auto-feedback scheme to constantly adjust the roller's vibrating frequency and amplitude resulting in a lower number of passes and greater compaction [1]. The equation that is used to describe the stiffness value (in MN/m) is:

$$k_b = \omega^2 \left[ m_d + \frac{m_0 e_0 \cos(\phi)}{Z_d} \right]$$

Where the stiffness value is related to the excitation frequency, drum mass, phase angle, vibration amplitude, and the eccentric moment of the unbalanced mass within the drum [3]. The soils stiffness value can also be computed from a vibration modulus, $E_{vib}$, or vice-versa as the values have been mathematically correlated[10].

### *Caterpillar Intelligent Compaction Equipment*
Caterpillar's system utilizes the standard GPS and accelerometer combination as well as a special slope sensor to monitor and improve compaction. This combination of equipment measures three variables call the compaction meter value (CMV), resonance meter value (RMV), and machine drive power (MDP). With these measurements, compaction can be measured at depths of up to 6ft (1.8 meters) [1]. The measurement of the compaction meter value (CMV) indirectly represents soil stiffness by associating it with the loss of drum contact with the soil during vibration [7].

Caterpillar's and Dynapac's compaction meter value (CMV) has been found to be a reliable method to correlate with soil stiffness. The accelerometers directly measure both the amplitude of the first harmonic frequency ($A_2$) as well as the fundamental frequency (A). The equation which describes this relationship is:

$$CMV = C\frac{A_2}{A}$$

Where C is a scalar constant used to bring the ratio of amplitudes up to a scaled value near 100. The resonance meter value (RMV) or bouncing value (BV, Dynapac) is described similarly:

$$RMV = BV = C\frac{A_{0.5}}{A}$$

Where $A_{0.5}$ represents the sub-harmonic acceleration amplitude or the amount of drum skips experienced on alternating vibratory cycles [3]. Caterpillar also uses a machine drive power (MDP) value which is calculates the amount of power needed to move the compactor. This equation is:

$$MDP = P_g - WV\left(\sin\alpha + \frac{a}{g}\right) - (mV + b)$$

Where $P_g$ is the gross power generated by the machine, W is the weight of the roller, V is the velocity, a is for acceleration, α is for the slope angle or inclination, g is the acceleration due to gravity, and finally m and b are internal loss coefficients which vary from machine to machine [3], [7].

### *Dynapac Intelligent Compaction Equipment*
Dynapac's proprietary system measures the dimensionless CMV value as well as a bouncing value (BV) which are the same as Caterpillar's measured value. Larger values of CMV and BV correlate to more compact or stiffer soils. If the BV value gets too large, the onboard computer will adjust the vibration amplitude to reduce drum bouncing or skipping and improve compactive effort [1].

### *Sakai Intelligent Compaction Equipment*
This manufacturer makes a fairly simple product which uses an accelerometer to measure a unitless value called the Sakai Compaction Control Value (CCV). This number represents the stiffness of the soil as it relates to how the vibratory roller interacts with the ground surface. Stiffer soils will cause the roller to experience a "jumping" or "skipping" effect. The computer measures this effect and correlates it to a stiffness value [1]. Sakai's compaction control

value (CCV) represents the ground stiffness through the use of the following equation:

$$CCV = \left[\frac{A_{0.5} + A_{1.5} + A_2 + A_{2.5} + A_3}{A_{0.5} + A}\right] x100$$

The different variables represent amplitudes of vibration at various frequencies [3].

# Intelligent Compaction Specifications and QA/QC

Prior to using an intelligent compaction enabled device, the onboard computer will require you to input project specifications in order to optimize the compaction. This may include information about lift thickness, soil type, gradation, and even weather conditions. Most intelligent compaction equipment also require you to enter a target value for compaction (stiffness or density) into the machine as well. Without the correct input of this kind of data, the computer may make incorrect calculations that could result in soils that don't meet quality standards.

No matter what intelligent compaction equipment you are using, there will still be a need for quality assurance and quality control. Prior to running the equipment, it should be calibrated using the manufacturer's recommended methods.

Quality assurance is typically achieved using one of 3 different methods. The first method includes taking point measurements on areas that have already been compacted and comparing them to the intelligent compactor's computer output and subsequently making any necessary changes to the inputs. The second method involves measuring the percent change in compaction between passes and comparing that to the computer output. Finally, the third option is correlating one or many point field tests to the intelligent compactor's reported value and then setting a target value to be achieved by the roller [2] based on these tests. Typical tests used for quality assurance/quality control include:

- Light Weight Deflectometer (LWD)
- Falling Weight Deflectometer (FWD)
- Briaud Compaction Device (BCD)
- Dynamic Seismic Pavement Analyzer (DSPA)
- Static Plate Loading Test (PLT)
- Dynamic Cone Penetrometer (DCP)
- Calibrated Nuclear Moisture-Density Gauge for soils and subbase (NG)
- GeoGauge (GEO)

Similar tests should be run after compaction is complete to verify that the machine did its' job in compacting the soil. Moisture testing is still extremely important and should be done using the standard moisture content measurement methods.

For quality assurance it's also important that adequate documentation be taken for verification purposes. The intelligent compaction enabled rollers typically have USB ports that can be used to retrieve data. The computer also generates geospatial data that can include everything from the vibrators frequency and amplitude at a particular location to the number of passes that were completed.

# Intelligent Compaction Correlations/Accuracy

Each type of intelligent compaction enabled roller produces a measured value that is directly proportional to the stiffness of the underlying soil layers. In most cases, the roller measured values produce an aggregate stiffness value that represents about 1 meter (~3.3ft) of the underlying soil material [7]. However, these stiffness values are often correlated to field or laboratory tests that only measure the aggregate stiffness of the top 0.2-0.3 meters (0.6 - 1.0ft) of soil. This difference in tests introduces uncertainty in the measured values that are observed at any given location.

Another area that may produce bias is soil moisture variation across the site being tested. Soils that have lower moisture content may have higher stiffness values. In clay soils low moisture content can result in significantly higher stiffness values - this is as a result of the clay structure [10]. This low moisture stiffness is a false value that, if not identified through moisture testing, can result in problems with the final product. Understanding these areas of uncertainty can help ascertain the limits of improvement for various soil types.

The FHWA reports that there is a linear correlation between the intelligent compactions measured value (either stiffness values or machine drive power) and the back-calculated layer stiffness values determined from measurements using the light weight deflectometer (LWD) or the falling weight deflectometer (FWD) for soils [3]. Reasonable correlations between the ICMV and California bearing ratio (CBR) values or dynamic cone penetrometer (DCP) tests were also observed for soils. With these measurements, small adjustments can be automatically made to the vibrating drum of the roller. The computer uses correlation equations to interpret the sensor's date and make the necessary changes to the machine.

More than 25 studies have been completed since the 1980's to look at the correlation between intelligent compaction measured values and the standard field tests typically used to monitor compaction in road construction [3]. The conclusions of these studies showed mixed results however most of them were positive. In addition, it must be noted that some manufacturer's measured values correlate better with standardized field tests than others.

In 2004 at a demonstration site in Edwards, Illinois,

Caterpillar's Machine Drive Power (MDP) showed good correlation with the dry unit weight ($R^2$=0.86) of sandy-silt soils however it did not correlate well with the Dynamic Cone Penetrometer (DCP) test which only had an $R^2$ value of 0.38 [3]. Further studies of Caterpillar's equipment in 2007 demonstrated better correlations between the machine drive power (MDP) and the Dynamic Cone Penetrometer (DCP). The correlation between machine drive power (MDP) and dry unit weight was revisited and this time the $R^2$ value was determined to be 0.92. The compaction meter value (CMV) had a poorer correlation with dry density only having an $R^2$ value 0.68 [8].

In a 2007 study, Ammann's stiffness value, $k_s$ show good correlation ($R^2$=0.80) with that of a plate load test (PLT) on a variety of soil types. However, poor correlations were found between the measured stiffness value and the light weight deflectometer (LWD), Dynamic Cone Penetrometer (DCP), Clegg Hammer, and Nuclear Gauge tests [3].

In another 2007 study of a roadway project in Kansas, Bomag's stiffness value ($E_{vib}$) was evaluated for correlation with several field tests. It was found that for the soils tested no good correlation could be made between the roller stiffness value and the DCP, FWD,

LWD, and Geogage tests. However, it was noted that the Bomag $E_{vib}$ stiffness value has had good correlations with plate load tests completed on many European projects [10]. However, the study did conclude that although a roller stiffness correlation could not be found, the intelligent compaction technology still provided a means to have uniform compaction and helped to located under-compacted areas that might have otherwise been missed.

Furthermore, it was found throughout the majority of the studies that the moisture content of a soil was a key factor in determining the final quality of the compacted material. Soil stiffness is very sensitive to changes in moisture content thus a good quality assurance/quality control  program is still necessary to ensure a good roadway is constructed.

# Intelligent Compaction Costs

A detailed economic analysis conducted by the Wyoming Depart of Transportation (WYDOT) found that construction costs with intelligent compaction enabled rollers were on the order of 54% less per lane mile for new roadways as compared to conventional compaction methods. There are also significant savings (upwards of $15,000 per lane mile) in annual maintenance throughout a roadway's life. This is achieved through the comprehensive monitoring of the soil, subbase, and pavement layers which ensures that 100% of the road structure gets compacted properly and consistently. The biggest savings were a result of the reduction in amount of quality control and quality assurance needed to manage the construction. Additionally, cost savings were realized by the 30% reduction in the construction time that was actually needed compact the roadway layers [2].

While construction and maintenance costs can be reduced, the initial investment in this technology can be very high. Retrofitting existing rollers can cost anywhere from $25,000 to $50,000 [5] whereas purchasing new equipment can cost as much as $500,000 dollars depending on the make and model of the machine.

Since this technology has yet to be adopted for use at a nationwide scale, it's difficult to predict whether contractor's bids will come in comparatively lower for new roadway projects. However, the cost to construct roadways with intelligent compaction technology should decrease in the future as more agencies require it and as more contracting companies invest in the equipment.

# Case History 1: Odessa, Delaware Test Site

In July 2008 the Burrice Borrow Put in Odessa, Delaware was selected for a pilot study of intelligent compaction technology. The project was to construct a 200ft long by 20ft wide embankment approximately 3ft high. The embankment was to be constructed in 5 lifts approximately 8 inches thick each using a poorly-graded sand with silt (SP-SM) which was found onsite [11], [13].

A Caterpillar CS56 smooth drum roller was used to compact each layer of the embankment. Compaction was performed using anywhere from 6 to 9 passes of the machine and was stopped when the computer's output show little improvement after the previous pass. The roller recorded the compaction meter value (CMV) as well as the machine drive power (MDP) throughout the compaction process.

Each layer was tested with 6 field tests at 19 different locations after the final compaction pass. Additional tests were also taken after the 1st, 2nd, 3rd, and 5th passes. The field tests include the GeoGauge, the light weight deflectometer (LWD), the nuclear density gauge, electronic density gauge, dynamic cone penetrometer

(DCP), and the falling weight deflectometer (FWD). Each test location also had a sample taken so that the soil moisture content could be determined and a standard proctor test could be performed [11].

The results showed poor correlation between the measured values and the field tests when all of the raw data was aggregated and used in the analysis. In other words, each field test value was correlated to dozens of data points that the roller computer recorded. The machine drive power measurement showed much better correlation with densities then the compaction meter value did [11], [12]. However, the best overall correlations seemed to occur with the dynamic cone penetrometer tests and the machine drive power (MDP) values having an $R^2$ value between 0.4 and 0.55 [13].

If the intelligent compaction measure values (CMV and MDP) were averaged for each lift and then compared to an average of the field tests, then the correlation became much greater. For the nuclear density gauge in comparison to the compaction meter value (CMV), an $R^2$ value was determined to be as high as 0.97 for a polynomial regression equation. For machine drive power (MDP) the $R^2$ value for this test was 0.98. This averaging of the data served to improve the correlations by reducing the bias introduced by the high volume of scatter or variable intelligent compaction measured values. The worst correlation was found to exist with the GeoGauge [13]. Table 1 below and 2 on the following page shows the results of these tests.

| Regression Analysis for Various Field Test Versus Compaction Meter Value (CMV) | | | |
|---|---|---|---|
| Test | Measurement | $R^2$ for All Soil Layers (Linear) | $R^2$ for All Soil Layers (Polynomial) |
| GeoGauge | Mpa | 0.20 | 0.63 |
| Light Weight Deflectometer 300mm | Mpa | 0.96 | 0.97 |
| Light Weight Deflectometer 200mm | Mpa | 0.94 | 0.95 |
| Dynamic Cone Penetrometer | mm/blow | 0.90 | 0.94 |
| Nuclear Density Gauge | $kN/m^3$ | 0.88 | 0.92 |

Table 1: Correlation Results for Compaction Meter Value (CMV) vs Field Tests, Average Results [13]

| Regression Analysis for Various Field Test Versus Machine Drive Power (MDP) | | | |
|---|---|---|---|
| Test | Measurement | $R^2$ for All Soil Layers (Linear) | $R^2$ for All Soil Layers (Polynomial) |
| GeoGauge | Mpa | 0.29 | 0.52 |
| Light Weight Deflectometer 300mm | Mpa | 0.90 | 0.95 |
| Light Weight Deflectometer 200mm | Mpa | 0.76 | 0.83 |
| Dynamic Cone Penetrometer | mm/blow | 0.90 | 0.92 |
| Nuclear Density Gauge | kN/m$^3$ | 0.91 | 0.91 |

Table 2: Correlation Results for Machine Drive Power (MDP) vs Field Tests, Average Results [13]

Overall, the results of this case study showed a strong correlation between average measured values from the intelligent compaction roller and the field test methods. However, point-to-point comparisons of field tests with singular measured values from the roller did not result in a good correlation. This was likely due to the amount of variability within the measured compaction meter value (CMV) and machine drive power (MDP) value.

# Case History 2: Atwater, Minnesota Test Site

In June of 2005 the Minnesota Department of Transportation (MNDOT) began a project to reconstruct an 11.9 mile section of TH12 from Willmar to Atwater, Minnesota. The project included removal of 30 inches of roadway structure and replacement with 6 inches of re-compacted subgrade, 14 inches of new granular subbase and 10 inches of hot mix asphalt. The project had 3 lanes that needed to be reconstructed.

For this job, the Bomag "Variocontrol" intelligent compactor was selected to roll to the subgrade and subbase. The roller was calibrated to compact the soil based on soil layers that were already deemed as passing in both compaction density and in uniformity. Target values for the vibration modulus were set using a draft specification (MNDOT) which was based upon testing various sites with the light weight deflectometer (LWD) and dynamic cone penetrometer (DCP).

A test location was selected for comparing the intelligent compaction measured values with field tests. The field tests included use of the GeoGage, light weight deflectometer, dynamic cone penetrometer, as well as proctor tests and sand cone tests. Soil moisture samples were also taken at each test location and tested in a laboratory oven as well. Once the Bomag roller was calibrated and the target stiffness values were determined, compaction of the soil layers at the test site began [14].

The results of the study showed that there was no real correlation between the vibration modulus (Bomag measured value) and any of the field tests. This essentially agreed with previous studies of Bomag's intelligent compaction technology. However, statistical analysis of each of the field tests as well as the intelligent compaction values showed that the modulus value measured from each type of test (including the Bomag values) had variations that were not statistically different from each other. In other words, the Bomag vibration modulus values were the same (statistically speaking) as the modulus values measured from the DCP and LWD. This result supports the proposition of using these modulus tests for setting target values for future compaction projects. Table 3 on the following page shows the results of the modulus values at 12 locations:

| Station | Lane | | Modulus Value | | | | | | |
|---------|------|-----|-----|-----|-----|-----|-----|-----|-----|
| | | DCP | DCP | | LWD | | | Geo Gauge | BoMag Avg |
| 65, 70, 75, 80 | A,B,C | DPI | Top 200 mm | 25cm | 50cm | 75cm | | |
| (ft) | | (mm/ blow) | (Mpa) | (Mpa) | (Mpa) | (Mpa) | (Mpa) | (Mpa) |
| 65 | A | 27 | 35 | 36 | 37 | 42 | 54 | 36 |
| 70 | A | 20 | 44 | 38 | 41 | 48 | 57 | 49 |
| 75 | A | 21 | 42 | 43 | 41 | 49 | 60 | 54 |
| 80 | A | 22 | 40 | 43 | 43 | 48 | 58 | 42 |
| 65 | B | 17 | 63 | 60 | 58 | 61 | 73 | 49 |
| 70 | B | 18 | 55 | 48 | 47 | 54 | 61 | 34 |
| 75 | B | 24 | 46 | 42 | 45 | 49 | 59 | 42 |
| 80 | B | 20 | 59 | 46 | 50 | 55 | 55 | 44 |
| 65 | C | 17 | 59 | 65 | 59 | 61 | 69 | 41 |
| 70 | C | 18 | 59 | 35 | 36 | 43 | 65 | 34 |
| 75 | C | 19 | 59 | 41 | 40 | 44 | 62 | 33 |
| 80 | C | 16 | 63 | 44 | 44 | 48 | 61 | 28 |

**Table 3: Modulus Results for Field Tests and Bomag Intelligent Compaction [14]**

Table 4 on the following page shows a statistical analysis of each of the field tests as well as the Bomag measured value:

| Device | Mean Modulus (Mpa) | Standard Deviation (Mpa) | Coefficient of Variation |
|---|---|---|---|
| DCP | 52 | 9.4 | 0.18 |
| LWD 25 | 45 | 8.5 | 0.19 |
| LWD 50 | 45 | 7 | 0.16 |
| LWD 75 | 50 | 6.1 | 0.12 |
| GEO | 61 | 5.3 | 0.09 |
| Bomag | 40 | 6.9 | 0.18 |

Table 4: Statistical Analysis of Modulus Values for Field Tests and Bomag Intelligent Compaction [14]

In addition to having no statistical differences with the other tests, the finished soil layers ended up meeting the standard specifications for compaction when the final sand cone and proctor tests values were compared. In fact, 100% of the sand cone tests that were completed onsite showed that the density was meeting or exceeding the specification when the lowest proctor density was assumed to be the optimum value (124.8lb/ft$^3$). About 67% of the sand cone tests showed good compaction when the highest Proctor value was selected as the target value instead (127.4lb/ft$^3$).

# Future Needs in Intelligent Compaction

Intelligent compaction is still very much an emerging technology. As such, the ability to find reliable information on the topic can be difficult. According to a survey conducted by the Wyoming Department of Transportation (WYDOT), fifty-one percent of the responders said that they obtained information about intelligent compaction from FHWA publications whereas only 11% obtained information from academic journals [2]. Citing only a relatively small amount of journal articles, intelligent compaction would certainly benefit from more research. More studies can help improve compaction or stiffness correlations while also helping to further validate the results of using intelligent compaction enabled technologies. In particular, more research is needed on the use of intelligent compaction on various types of soils as well as correlating the measured values to field/laboratory modulus tests [4].

Many states in America still have not even had their first intelligent compaction project. In fact, the State of Arizona just had its first pilot project in early 2015 [5]. Fortunately, the Federal Highway Administration has a grant program which is working to encourage the widespread use of this technology. Contractor's also have to make significant investments in new or modified equipment as well.

Moreover, there is a need to further develop standards for the use of intelligent compaction in roadway and foundation design and construction. Currently there is no nationwide specification for the use of intelligent compaction however, the Federal Highway Administration (FHWA) has prepared many documents and guidelines on the subject. Currently, only 18 states have drafted or adopted intelligent compaction quality control and assurance specifications. For now, only 8 of those states have specifications for soil compaction [2]. The other states have focused their efforts on pavements.

In addition to these needs, it is become apparent that further improvements in intelligent compaction software is needed. This comes both in the areas of stiffness correlations as well as the data management and reporting methods. With additional research, an improved (or even universal) model can be found to relate the machine's measured values with a true stiffness value that can be validated in both the field and in the lab [3]. Standardization of the intelligent compaction data is needed to help improve compatibility of the reported data between contractors, agencies, and academia. Reporting standards also need to be developed to ensure consistent information is generated and stored during and after construction projects.

# Conclusions

Touting lower costs, improved compaction efficiency, and a higher quality pavement structure, intelligent compaction is a young technology that has significant potential to change the way that roads are built in the future. Intelligent compaction utilizes modern technological tools to measure compactive effort in real time from within the roller. The onboard computer uses an auto feedback system to monitor and make changes to the drums frequency, amplitude and even rolling speed to ensure optimum compaction efficiency. Intelligent compaction generates vast amounts of data that can be used to assess compaction consistency and uniformity between lifts for 100% of the project site.

When paired with modulus based laboratory and field tests, intelligent compaction systems have been shown to be a reliable way to measure compaction and assure roadway quality and longevity. Several studies have shown good correlations between the intelligent compaction machine's measured values and field tests. In the cases where correlations could not be established, it was generally found that the measured values were not statistically different than other field tests showing that this method is at least as reliable for measuring compaction as the other methods are.

In addition to the stiffness of the pavement, subbase, or subgrade section, improved roadway quality is also achieved by more uniform and consistent compaction. This can only be attained with intelligent compaction equipment.

However, there is still plenty of work to be done before full scale national implementation of intelligent compaction technology will be achieved. Standardization of the intelligent compaction methodologies and the generated data needs to be undertaken followed by the development of a specification for how this technology should be utilized and under what conditions. More research is needed to characterize target modulus values (based on one field/lab tests) for a variety of soil types. And finally, a shift within the engineering community as well as state agencies will need to occur that steers current practices away from testing and measuring soil density and towards the use and measurement of the soil's modulus instead.

# References

[1]   U.S. Department of Transportation, Federal Highway Administration. (2010). Intelligent Compaction for Soils and Subbase Materials (FHWA Tech Brief DTF61-07-C-R0032). Indianapolis, IN. Retrieved from

[2] Savan, Christopher., Weng Ng, Kam., Ksaibati, Khaled. "Implementation of Intelligent Compaction Technologies for Road Construction in Wyoming." Department of Civil and Architectural Engineering, University of Wyoming. March 2015. Retrieved from

[3] U.S. Department of Transportation, Federal Highway Administration. (2011). Accelerated Implementation of Intelligent Compaction Technology for Embankment Subgrade Soils, Aggregate Base, and Asphalt Pavement Materials. (Publication FHWA-IF-12-002). Retrieved from

[4] Briaud, Jean-Louis. "Intelligent Compaction: An Overview." Department of Civil Engineering. Texas A&M University. March 25th, 2005. Retrieved from

[5] Arizona Department of Transportation. "ADOT to test innovative construction technology to extend life of roadways." October 17, 2014. Retrieved from

[6] Vennapusa, Pavana Kumar Reddy, "Investigation of roller-integrated compaction monitoring and in-situ testing technologies for characterization of pavement foundation layers" (2008).Graduate Thesis and Dissertations. Paper 11190.

[7] Heersink, Daniel K., and Reinhard Furrer. "Spatial analysis of modern soil compaction roller measurement values." *Procedia Environmental Sciences* 7 (2011): 8-13. Retrieved from

[8] White, David., Thompson, Mark, "Field Calibration and Spatial Analysis of Compaction-Monitoring Technology Measurements." *Transportation Research Record Journal of the Transportation Research Board.* January 2007. Retrieved from

[9] Vennapusa, P., White, D. J., and Morris, M. (2009). "Geostatistical analysis of spatially referenced roller-integrated compaction measurements," Journal of Geotechnical and Geoenvironmental Engineering, ASCE, 136(6), 813-822.
http://lib.dr.iastate.edu/cgi/viewcontent.cgi?article=1000&context=ccee_pubs

[10] Rahman, Farhana, et al. "Intelligent compaction control of highway embankment soil." *86th Annual Meeting of the Transportation Research Board.* 2007. Retrieved from
http://citeseerx.ist.psu.edu/viewdoc/download?doi=10.1.1.668.9113&rep=rep1&type=pdf

[11] Tehrani, F. S., and Christopher L. Meehan. "Continuous Compaction Control: Preliminary Data from a Delaware Case Study." *Proc., Eighth International Conference on the Bearing Capacity of Roads, Railways and Airfields (BCR2A'09), Champaign, IL, June.* 2009. Retrieved from:
https://www.researchgate.net/profile/Faraz_Tehrani/publication/273022071_Continuous_compaction_control_preliminary_data_from_a_Delaware_case_study/links/559a33400c5d8193658ee.pdf

[12] Meehan, Christopher L. "An introduction to Continuous Compaction Control Systems" *Presentation at DelDOT Winter Workshop.* February 18th, 2010.

[13] Meehan, Christopher L., Cacciola, Daniel V. "Using Continuous Compaction Control Systems within an Earthwork Compaction Specification Framework." Department of Civil and Environmental Engineering, Delaware Center for Transportation, University of Delaware. October, 2013. Retrieved from
https://sites.udel.edu/dct/files/2013/10/Rpt.-236-Continuous-Compaction-Control-1avxgl3.pdf

[14] Carmargo, F., Larsen, B., Chadbourn, B., Roberson, R., Siekmeier. "Intelligent Compaction: A Case History." *54th Annual University of Minnesota Geotechnical Conference.* February 17th, 2006.